Citizen George Coloring Book Volume 2

Authored by Michael Dow, RN, MS, MHA, MSM

Drawings by Angel Kincaid

Citizen George Coloring Book Volume 2

Copyright © 2026 by Dow Creative Enterprises, LLC

All rights reserved

First Edition

ISBN 978-1-968690-06-9

Printed by Lulu.com

Published by Dow Creative Enterprises®

Dow Creative Enterprises, LLC

PO Box 15357

Tucson, AZ 85708

Library of Congress Control Number:

Dow Creative Enterprises® is a federally registered trademark with the United States Patent and Trademark Office.

Help Civilization Reach Its Potential® is a federally registered trademark with the United States Patent and Trademark Office.

Table of Contents

Dedication

This book series is dedicated first to Florence Nightingale, the Nurse Pioneer in the 1800s. She helped create nursing into a profession and the healthcare system as a whole owes her gratitude. Second, I'd like to dedicate this book series to Dr. Jean Watson, who developed the Science of Human Caring for nurses, but it can also be applied to all healthcare professionals. May all nurses continue to develop stronger trusting and helpful relationships with all patients as they bring patients into complete well-being. I'd also like to dedicate this book series to the staff at the College of Nursing at the University of Arizona for their outstanding instruction. Thank you.

Michael

Background

Stephen and his Uncle George are going to the nearby lake to walk around and talk. George likes to take his nephews and nieces on walks to talk about life and living well.

“Thank you so much for spending time with me today, Uncle George. I know how busy you are,” says Stephen.

“Spending quality time with loved ones is part of a happy life with contentment. Being happy doesn’t mean you are always joyful though,” says George.

“This lake has always been special to me. I still wonder how this Earth got so much water on it. Always stay curious and maybe one day you’ll discover the answers to your questions,” says George.

"Living a dignified life is important so that you have a sense of pride in who you are as a person. That's what our lesson today will be about. Dignity means being respected, honored, and valued regardless of your circumstance, actions, or status," says George as he throws food pellets into the water for the ducks.

“The proverb that I want us to focus on today is ‘When in company, put not your hands to any part of the body not usually discovered.’ The parts not usually discovered are mainly referring to the groin area or your private parts,” says George.

“We should avoid touching our privates in public as well as our rear end. If it starts to itch, then we should go to the restroom to take care of ourselves,” says George.

“Now, I want you to think of some things that can be done to live a dignified life,” says George. They spend a couple of minutes in silence while Stephen thinks about it. Stephen also throws some food pellets into the water.

“To be respected no matter your actions might involve slowing down your speech so that you speak confidently and clearly. Talking with others is important so we spend a lot of our lives communicating with others,” says Stephen.

“You’re exactly right. Talking with others is a skill that we need to develop,” says George.

“Regarding speech, if you pause before answering someone’s question, then it shows a measure of respect that you are carefully thinking about their question and taking time to thoughtfully respond to someone. It’s ok to pause for 3 seconds before answering sometimes,” says George.

“Sometimes I hear people say filler words a lot like ‘um, ‘like’, and ‘you know.’ It can be distracting to hear a lot of filler words and gives the feeling that the person is not being careful with their words. Speaking dignified should include avoiding filler words,” says Stephen.

“Concise and clear communication is best,” says George.

"How someone stands or sits as they face you can be factor in living dignified. A person should keep their posture open and relaxed and avoid showing guardedness. Limit how much you cross your arms in front of your chest," says George.

“Remember our last conversation about listening to others and respecting those around us. We talked about how having steady eye contact is important for most cultures. Living dignified might also include good eye contact,” says Stephen.

“Great point,” says George.

“When a person moves, they should move with purpose. Living with purpose and doing everything with purpose shows you value your life and appreciate the universe for its gift. Remember, life is the greatest gift that you will ever be given,” says George.

“Since we wear clothes when we are around others, a person should have clean clothes to live dignified,” says Stephen.

“Some people do dirty jobs so they should bathe once they get home and dress in casual clean clothes to end their day. Also, remember that living clean will help you feel clean,” says George.

“Dressing simply can be important to avoid getting too much attention to your wardrobe. Let’s avoid arrogance. Let’s dress very, very nice when everyone else does at a special event such as a wedding,” says George.

“I feel a person should use more I statements instead of You statements. Sometimes using You statements can be accusatory and that may not be what you mean. Explaining how you feel instead of telling another person how they feel or should feel would be best,” says Stephen.

“Great idea,” says George.

"Being kind to others not only helps out society, but it helps you feel good about yourself and feel dignified. You never know what someone else is going through so kindness can help most relationships," says George.

“I think it’s important to listen more and talk less, but balance the conversation so plan to talk more if the other person doesn’t want to talk,” says Stephen.

“I like it,” says George.

“Something very important with living a dignified life is accepting blame for your mistakes and owning them. It helps you live in your truth of who you know yourself to be and helps others understand boundaries that you and they may have,” says George.

“I think respecting other people’s boundaries is important since that could help others respect your own boundaries,” says Stephen.

“Excellent,” says George.

"A person should be authentic and true to who they are. Being yourself is important to live dignified since valuing who you are means knowing who you are. Know thyself. Living true to yourself encourages others to live true to themself," says George.

“I think being grateful to the universe for this life and what you’ve been given is helpful,” says Stephen.

“I believe the universe does appreciate our appreciation,” says George.

“It’s important to keep learning and growing as a person which shows respect to the complexity of our brain for its abilities and capabilities to understand this life. If you never fail, then you’re not even trying,” says George.

“Florence Nightingale said ‘Let us never consider ourselves finished nurses, we must be learning all of our lives.’ These are good words to apply to all people since we are all on a journey and continuous improvement should be a goal,” says George.

“Also, always remember that a dignified life goes both ways. What I mean is that you should try to live in dignity as well as help others live in dignity,” says George.

"Thanks Uncle George for taking time to spend with me," says Stephen.

"You're welcome," says George as they throw some more food to the ducks.

George and Stephen then leave the lake and walk back to the house. Stephen thinks about what was said throughout the day and practices these ideas with his mom and dad.

References

Brown, L. (2024). 8 ways to carry yourself with dignity and respect. Retrieved from: https://theconsideredman.org/8-ways-to-carry-yourself-with-dignity-and-respect/

Miller, L. (2025). 9 ways to be more dignified and respected in how you speak and carry yourself. Retrieved from: https://cottonwoodpsychology.com/blog/9-ways-to-be-more-dignified-and-respected-in-how-you-speak-and-carry-yourself/

Journal

What page number was your favorite drawing?

What were three things you learned?

Do you have any new questions about life or the body from reading this book?

What do you plan to do differently in your life after reading this book?

About the Illustrator

Angel Kincaid is an artist who studied at, Art Instruction School, American National, and the University of Kentucky for various kinds of art. She is a forensic artist and is also trained in illustration as well as classical painting. She lives in Lexington, Kentucky with her husband and lil' dog. Originally from Utah, she found she loved art at a young age and has studied and done art her whole life. She is currently working on her PhD in forensics/forensic art at the University of Kentucky and looking forward to doing more with her art career.

About the Author

Michael is married to Perla, and they have three children. Michael served in the US Air Force between 2002 and 2010 as an Electronic Warfare Officer on the EC-130H Compass Call and deployed 6 times in the Global War on Terror. Michael then served 8 years as an Army Wounded Warrior Advocate. Michael used his GI bill to go to nursing school and works as an RN at an inpatient psychiatric hospital in Tucson, AZ. Michael enjoys listening to Beethoven and reading a lot of news.

Michael's college education:

B.A. in Psychology from Auburn University,

B.S. in Biology from the University of Alabama at Birmingham,

M.S. in Management from Troy University,

Master in Health Administration from the University of Phoenix,

M.S. from the University of Arizona through the accelerated Master's Entry to the Profession of Nursing program

Other Books by Dow Creative Enterprises®

Nurse Florence®

www.nurseflorence.org

Nurse Dorothea®

www.nursedorothea.com

Citizen George

www.citizengeorge.com

Visit www.DowCreativeEnterprises.com for more information

www.ingramcontent.com/pod-product-compliance
Lightning Source LLC
LaVergne TN
LVHW011051110826
845149LV00015B/3450

* 9 7 8 1 9 6 8 6 9 0 0 6 9 *